Do·It·Yourself
HOME
PLUMBING

Blandford Press
Poole·Dorset

Series Editor: Mike Lawrence
Author: Roger Bisby

Editorial production by Keith Faulkner Publishing Limited

© Aura Books Limited 1985

This edition first published by
Blandford Press 1985
Link House
West Street
Poole Dorset

Printed and bound by Henri Proost, Turnhout, Belgium

Contents

Understanding the plumbing

The apparent jumble of pipes and stopcocks around your home may at first glance seem confusing. But with a little examination and careful tracing you should soon be able to find out what they do and how they all work.

Start with the cold supply by finding the internal stop cock; this controls the water coming into your house from the mains. Usually, it's at low level near the floor. It should turn on and off fairly easily; if it doesn't turn the water off completely it will need servicing (see page 27).

Trace the route of the pipe to your cold water tap and beyond to the cold storage tank at the top of the house. A ball valve with a float shuts the water off when the water level is a few inches from the top of the tank.

On some systems the rising main supplies other taps and fitments such as the WC. You can find out if this is so by checking them with the mains turned off. If they are still working then you can assume they are fed with an indirect low pressure supply from the cold water tank.

Trace the pipes coming out from the bottom of the cold water tank. Somewhere along these pipes there should be gate valves or stopcocks. These may be near the tank or in the airing cupboard. One valve controls the cold supply to the bathroom and WC – check it works by turning on the cold tap in the bath and then turning off the gate valve on the supply pipe. Once you've discovered which valve controls the cold supply tie a label on it so you can recognize it in the future.

The remaining valve controls the cold supply to the hot water cylinder –

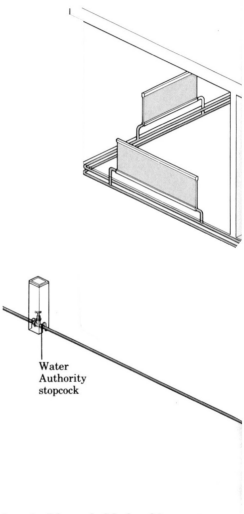

Water
Authority
stopcock

A typical household plumbing system, showing rising main, hot and cold supply and drainage.

this can be checked by turning on any hot tap. The water should stop after a few seconds with the gatevalve turned off. The fact that a valve on the cold supply turns off the water to the hot taps is a mystery to many people. By following the pipe a little further down its route you will soon discover how this is so.

The pipe enters the bottom of the hot water cylinder to feed the cylinder with cold water. As the water is

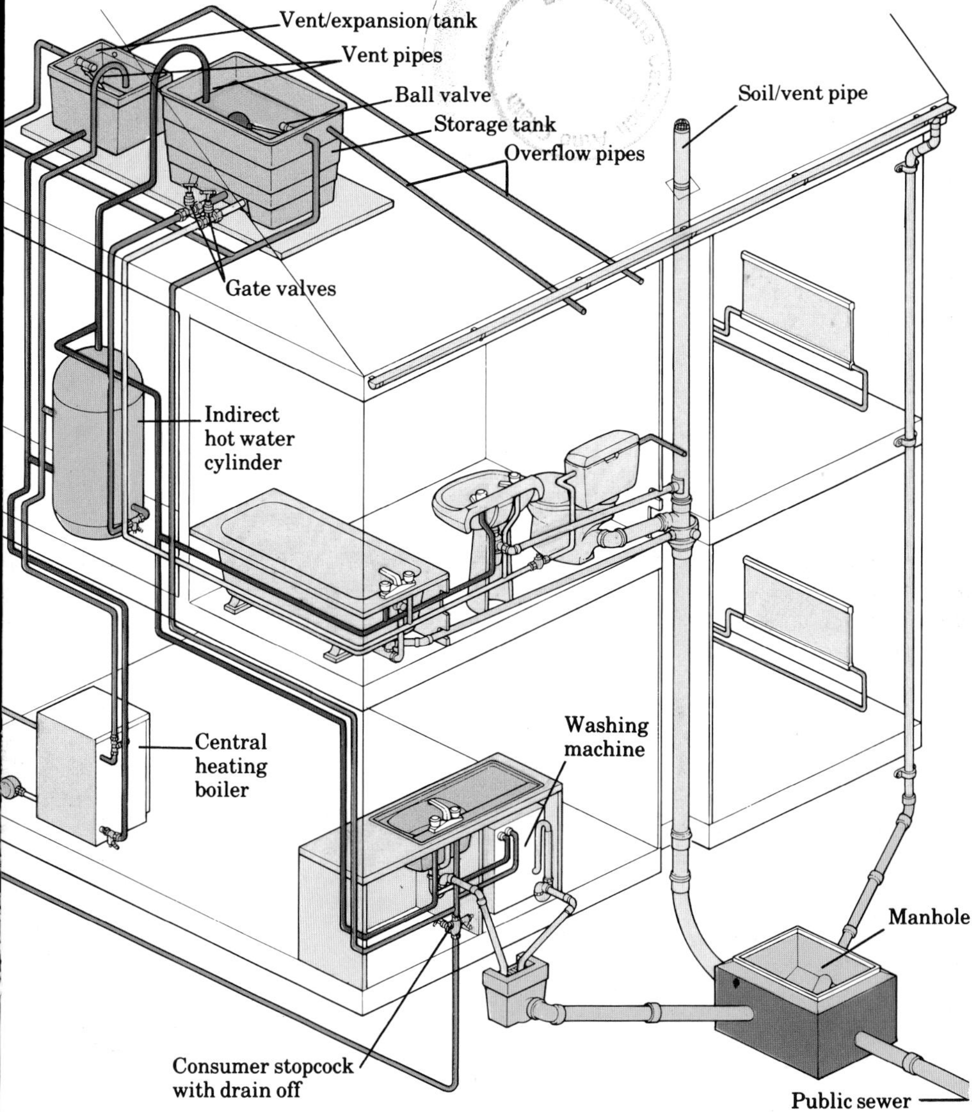

Vent/expansion tank
Vent pipes
Ball valve
Storage tank
Overflow pipes
Soil/vent pipe
Gate valves
Indirect hot water cylinder
Central heating boiler
Washing machine
Manhole
Public sewer
Consumer stopcock with drain off

heated it rises to the top of the cylinder where it waits to flow along the pipes as soon as the hot tap is turned on. The pressure to push the water along the pipes is supplied by the cold water tank sitting several feet above the height of the taps. The higher the cold tank the greater the 'head' pressure. This is particularly important where showers are concerned. If you were to place a shower spray at the same height as the cold water tank no water would come out at all because the water would find its own level in the tank and pipe. This is exactly what happens in the vent pipe which rises from the top of the cylinder and hooks over the cold tank.

Stop valves or gate valves control all hot taps and bathroom cold supplies

It's important to know the control points in your plumbing system. Once you've located them, tie an identifying label to each one.

Isolating valve on WC supply

Rising main stopcock with integral drain off

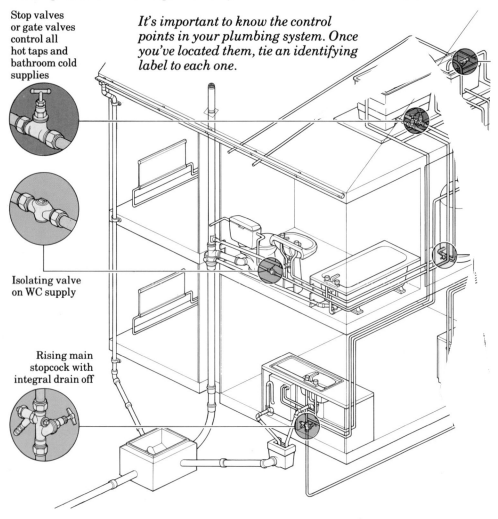

By using this simple principle a natural pressure vent is provided during heating up of the water but a constant balanced pressure means water will flow easily along the hot pipes.

Other less common systems than the one shown are found in many homes. Most of these use some form of

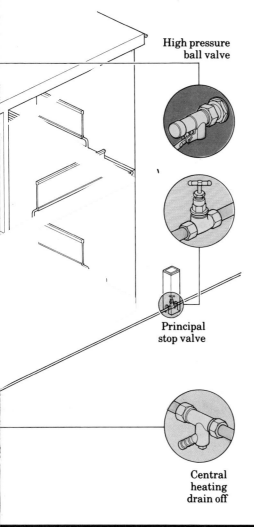

High pressure ball valve

Principal stop valve

Central heating drain off

instantaneous heater often supplied direct from the mains. To turn off the water to the hot taps the cold supply to the heater is turned off; this means that every tap in the house is controlled by one mains stopcock.

Having seen how the water is brought into your home it is equally important to see how it is taken away.

Several varieties of waste system exist, depending upon the age of your property. The modern tendency is towards a neat simple one pipe system that sends used bath, basin and WC water into a single soil and vent pipe known as the soil stack. This is open to the atmosphere at a high level to prevent pressure or vacuum being built up when wastes are discharging. The water then falls to a manhole and is taken away through the drains to the main sewer. In some older properties the drains are shared with a neighbour and are often maintained by the local authority as part of the public sewer system. It's worth finding out if this applies to your system since any blockage may be cleared free of charge.

The kitchen sink waste usually enters the drains via a trapped gully buried in the ground outside the house. This is in fact simply a large U-bend which stops smells escaping.

On many houses the rainwater pipes discharge into a completely separate system from the foul water drains. Sometimes surface water and foul water drains are laid side by side or there may be a soakaway buried in the garden that takes the water from the gutters. It's important not to discharge a washing machine or sink waste into a surface water drain.

Basic techniques

Mastering a few basic plumbing skills will enable you to carry out all the projects described in this book. The range of quality fittings available in Texas stores have been chosen to bring plumbing well within the grasp of the novice. Especially helpful are the kits containing all the parts necessary to carry out the more popular jobs, such as installing a basin or connecting a washing machine.

With the range of tools shown here, you should be able to tackle most plumbing jobs.

Gas blowlamp

Self-cleaning flux

Fire resistant mat

Ordinary flux

Solder

Round and half round files

Pipe cutter

Trimming knife

Open ended adjustable spanners

Internal bending spring

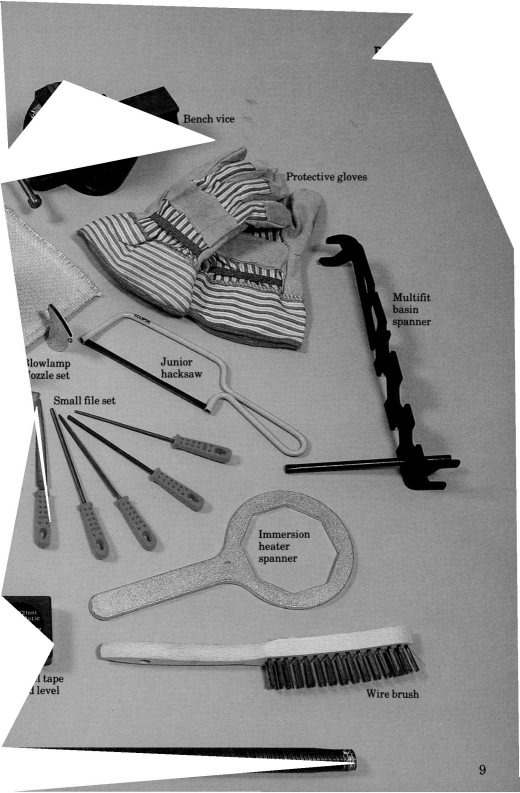

Bench vice

Protective gloves

Multifit basin spanner

Blowlamp Nozzle set

Junior hacksaw

Small file set

Immersion heater spanner

tape level

Wire brush

9

CUTTING COPPER PIPES

Cutting copper pipe can be done in two ways. You can use a junior hacksaw for small jobs – with practice you should be able to make a neat cut keeping the pipe ends square. Use a small flat file to remove any burrs.

If you're planning to do a lot of cutting you can invest in a small pipe cutter. This ensures that the cut is square and neat. Place the cutter on the tube and gently tighten the wheel onto the pipe, revolving the cutter around the pipe as you tighten. You will very soon see a cutting line appear all the way round the pipe and it will fall neatly in two. Use the pointed reamer on the cutter to re-

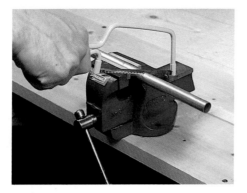

With a little practice you will be able to cut your pipes square using a junior hacksaw.

After cutting the pipe, the burr must be removed with a file, flat for the outside and round for the inside.

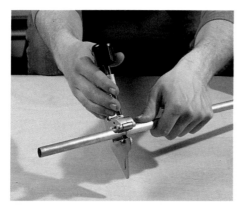

If you plan to do a lot of plumbing jobs in the home, invest in a small, hand pipecutter. It will cut neatly and

cleanly without leaving an external burr, and has a reamer for removing the internal burr.

move the internal burr from the pipe.

BENDING COPPER PIPES

To save on fittings and make a neat job you may wish to bend copper pipe. This can be done by using an internal bending spring. Thread the spring through the pipe until it is positioned at the point you wish to make the bend. Placing the pipe across your knee, gently pull the bend around to the desired angle. Bend the pipe a little further than you require and then bend it back to the correct angle – this will help free the spring for easy removal.

For difficult bends, especially on larger bore pipes such as 22mm, the job can be made much easier by annealing the pipe.

Take the copper pipe and heat the section that is to be bent either with a blowlamp or over a gas stove. Hold the pipe using a rag until it glows cherry red on the bending area, then allow the pipe to cool thoroughly.

The result will be a far softer section of pipe which can be bent to the required angle in the normal manner using a spring.

Mark the pipe clearly showing the position of your intended bend.

For difficult bends and larger bore pipes, the copper must be annealed first. Just heat it to cherry red and then let it cool.

Insert the oiled bending spring to the correct position. A length of string will help you to remove it.

Use your knee to bend the pipe to the desired angle.

JOINING COPPER PIPES WITH COMPRESSION FITTINGS

You can join copper pipes quickly and easily using compression fittings. A wide range is available for use on hot and cold systems capable of withstanding great pressure, so you can have complete confidence in a properly made joint.

Compression fittings come ready-assembled. Undo the nuts and place them on the pipes to be joined, then slide on the rings. Place the fitting on and align the pipes so they enter squarely. Smear a little jointing compound around the mouths of the fitting and tighten home the nuts trapping the rings in position – make

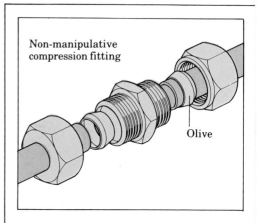

Non-manipulative compression fitting

Olive

There are two types of compression joint, manipulative and non-manipulative. Always use jointing compound smeared around the mouth

SAFETY WITH BLOWLAMPS

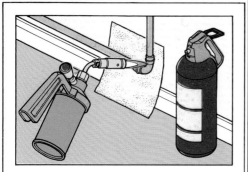

Protect flammable surfaces and decorations with a glass fibre mat. Keep a container of water or better still a small fire extinguisher handy just in case. After making a joint check that nothing is smouldering particularly in dusty crevices. Finally, go back and check the area with in half an hour.

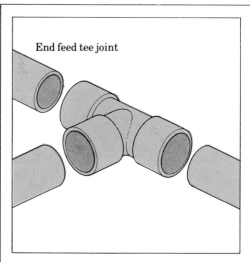

End feed tee joint

End feed joints rely on capillary action to ensure an even distribution of solder in the joint. About 20mm of solder per joint should be sufficient.

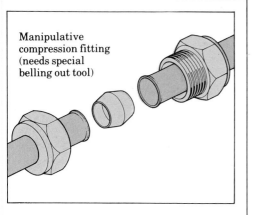

Manipulative compression fitting (needs special belling out tool)

of the fitting and pipe to ensure that the joint is watertight. Manipulative compression fittings require a special tool to bell out the ends of the pipe.

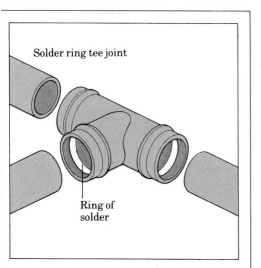

Solder ring tee joint

Ring of solder

Pre-soldered joints have a ring of solder within the fitting. Assemble and heat until an unbroken line of solder appears around the rim.

sure the pipes are firmly in the fittings as you tighten. There's no need to overtighten the nuts; about a turn past hand tight will be plenty, anything more than this will distort the pipe and ring.

As an alternative to jointing compound you can use PTFE tape around the thread of any screw fitting.

JOINING COPPER PIPES WITH CAPILLARY FITTINGS

Capillary fittings are made with a blowlamp, they are neater and cheaper than compression fittings and provide an equally good joint on all copper pipework. They are useful where space is restricted but because they need heat they're not suitable near plastic fitments.

It's essential that the pipe is empty of water – even the slightest drop will prevent the solder running evenly. If you're using self-cleaning flux there's no need to polish the pipe provided it is untarnished and free of grease. With conventional fluxes you must make sure the pipe ends and the inside of the fittings are shined bright with steel wool.

Unless using self-cleaning flux always burnish the pipe end and fitting with steel wool.

Apply a coating of flux evenly onto the pipe end to ensure that the solder bonds to the copper.

Smear a little flux evenly on the pipe ends and insert them into the fitting so they reach the internal stops. Using a blowlamp gently heat the fitting. If you're using solder ring fittings the solder will eventually appear at the mouth of the fitting; move the blowlamp on to the other end of the fitting until this does the same. Immediately you see a continuous ring of solder around the mouth of the fitting take the heat away.

You can heat a pre-soldered joint with a blowlamp or even over a gas stove.

End feed capillary fittings are cheaper than solder ring fittings but you must feed the solder in yourself. This means you'll have to buy a reel of solder which is only worthwhile if you're making several joints. The procedure is the same except that when the joint is heated you'll need to touch the mouth of the fitting with the solder. The heat will draw the melted solder into the fitting within seconds. It works equally well whatever position the joint is in since capillary action draws solder up as

When soldering an end feed joint, about 20mm of solder is sufficient.

well as down.

To ensure you feed enough solder into the fitting bend the end of the wire into zig zags approximately 20mm long and use a section on each end of the joint.

For soldering plumbing joints a gas blowlamp is safe and easy to use. As a precaution always keep a bucket of water or a small fire extinguisher close at hand, and use thick gloves to protect your hands.

Basic techniques

*At every TEXAS HOMECARE store
there is a vast range of compression
and capillary fittings to suit any
possible home plumbing needs.*

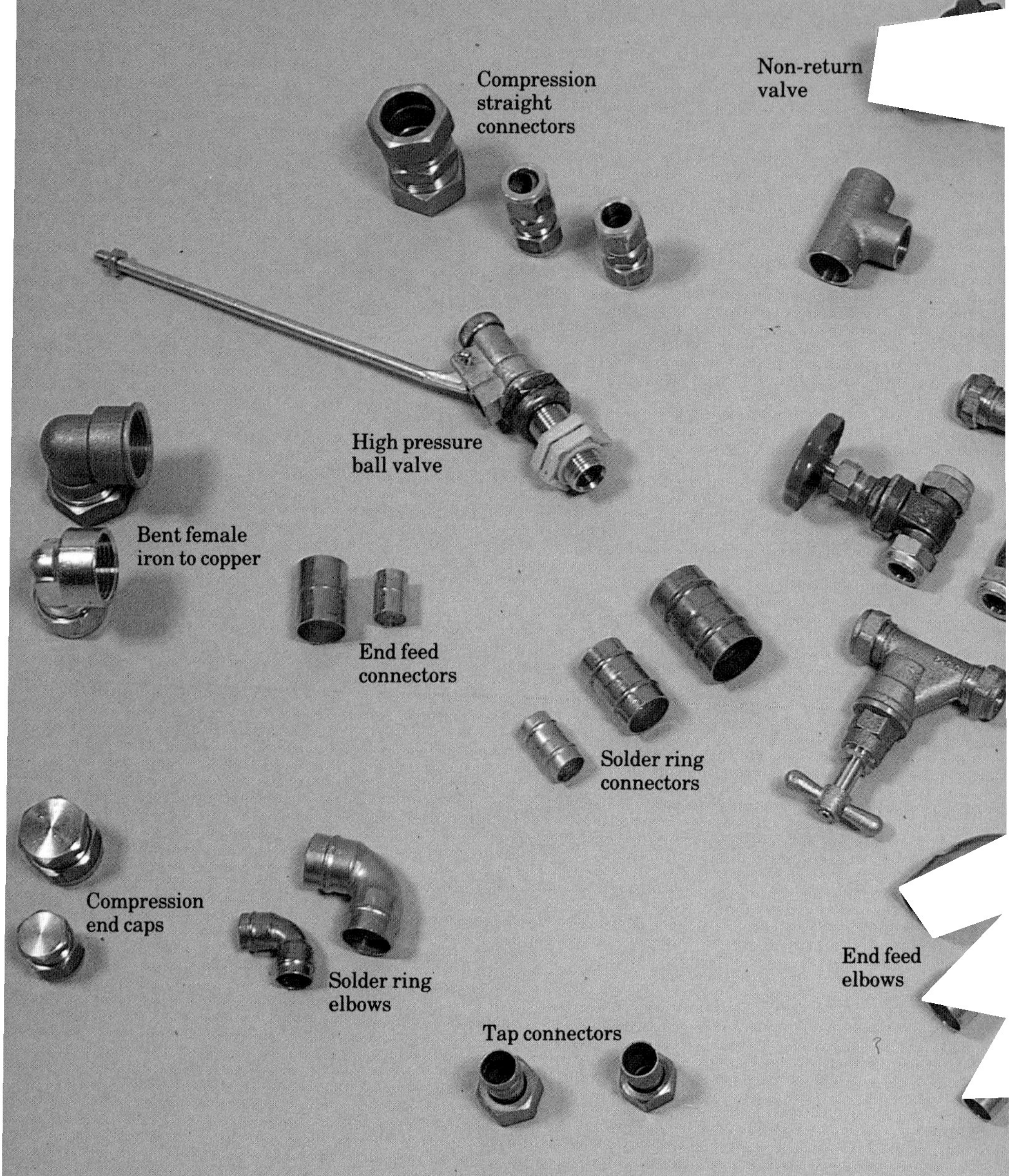

Compression
straight
connectors

Non-return
valve

High pressure
ball valve

Bent female
iron to copper

End feed
connectors

Solder ring
connectors

Compression
end caps

Solder ring
elbows

End feed
elbows

Tap connectors

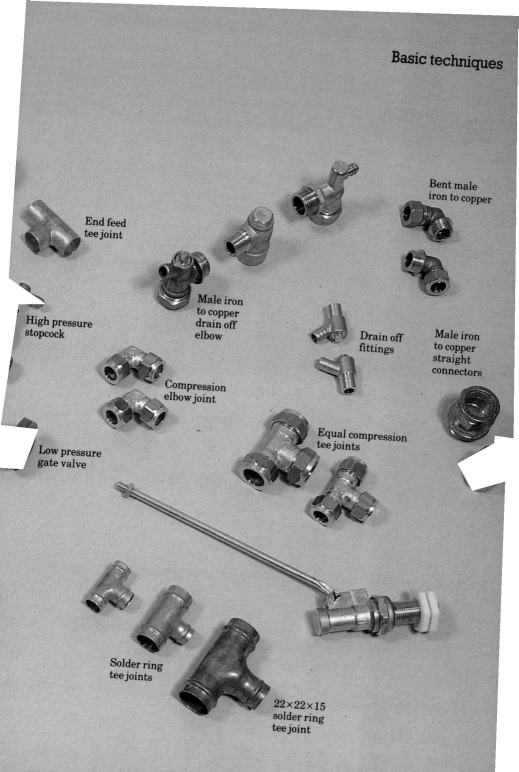

End feed
tee joint

Bent male
iron to copper

High pressure
stopcock

Male iron
to copper
drain off
elbow

Drain off
fittings

Male iron
to copper
straight
connectors

Compression
elbow joint

Low pressure
gate valve

Equal compression
tee joints

Solder ring
tee joints

22×22×15
solder ring
tee joint

You can use an ordinary fine toothed saw to cut plastic pipes.

Mark the pipe clearly, allowing about 10mm for expansion.

You can assemble the joint more easily if you lubricate it first, with a small amount of silicone grease.

RUNNING PUSH FIT WASTE SYSTEMS

A comprehensive range of plastic fittings and pipes makes quick work of any waste system.

A few simple techniques will help you achieve a first class job every time. Using the guide below plan out your intended waste system and make a note of the fittings and pipe you'll need.

Sinks and baths use a 40mm (14in) bore system whilst basins use 32mm (1¾in) except for runs of over two metres where the pipe size should be increased to 40mm. Bends are available to change direction and a 'T'junction allows you to join two runs together. If you're connecting straight into a soil stack using a connector boss you'll need to cut a hole in the stack. Never join two waste runs from different fitments together when connecting to a stack.

You can cut the pipe using an ordinary fine tooth saw – smooth off with sandpaper to avoid sharp edges. Use a pipe clip as a cutting guide to give you a square cut every time.

When connecting push fittings smear a little washing-up liquid on, to ease the seal. Draw the pipe back out 10mm to allow for expansion.

Try to maintain a gentle fall – about 1 in 20 – and support the pipe with clips every 500mm.

RUNNING PIPES

Whatever plumbing job you're doing the end result will be more pleasing if you run the pipework in a neat and orderly fashion. All too often the hallmark of an amateur job is untidy pipework that hasn't been thought out. You can avoid this by planning ahead, keeping pipes out of view, clipping them neatly at regular intervals making sure the clips are parallel and level, bending pipes where possible to avoid too many fittings, running the pipes alongside features such as skirtings, architraves and existing pipes.

Where you need to run pipes under floorboards you'll probably have to notch joists. It's important to bear in mind that the joists are structural and should not be unduly weakened. Golden rules for notching joists are:

● only notch in the first third of the span i.e. near to the wall.

● never notch deeper than one sixth of the depth.

● never notch the underside of a joist

● always notch beneath the centre of a floorboard to avoid damage when nailing back down and leave enough support for the boards either side of the pipe.

The problem of air locks sometimes occurs in low pressure pipework because the pipe rises to a high spot where the air becomes trapped; this can usually be avoided if the pipes rise towards the taps so any air is naturally vented.

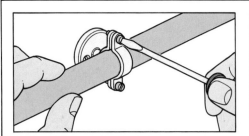

Fit pipe clips at regular intervals. This can eliminate vibration when the pipes are in use.

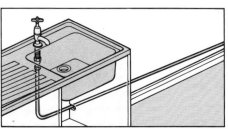

Always ensure the pipes run slightly up towards the taps so that any air is naturally vented.

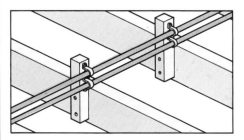

Pipework in the loft must be supported. Clip the pipes to battens screwed to joists or rafters.

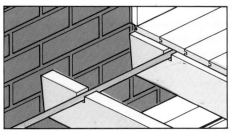

Follow the golden rules for notching joists, and remember that they are a structural support.

Basic techniques

Fitting waste systems and overflows is now a simple matter. These plastic pipes and fittings are quick and easy to assemble.

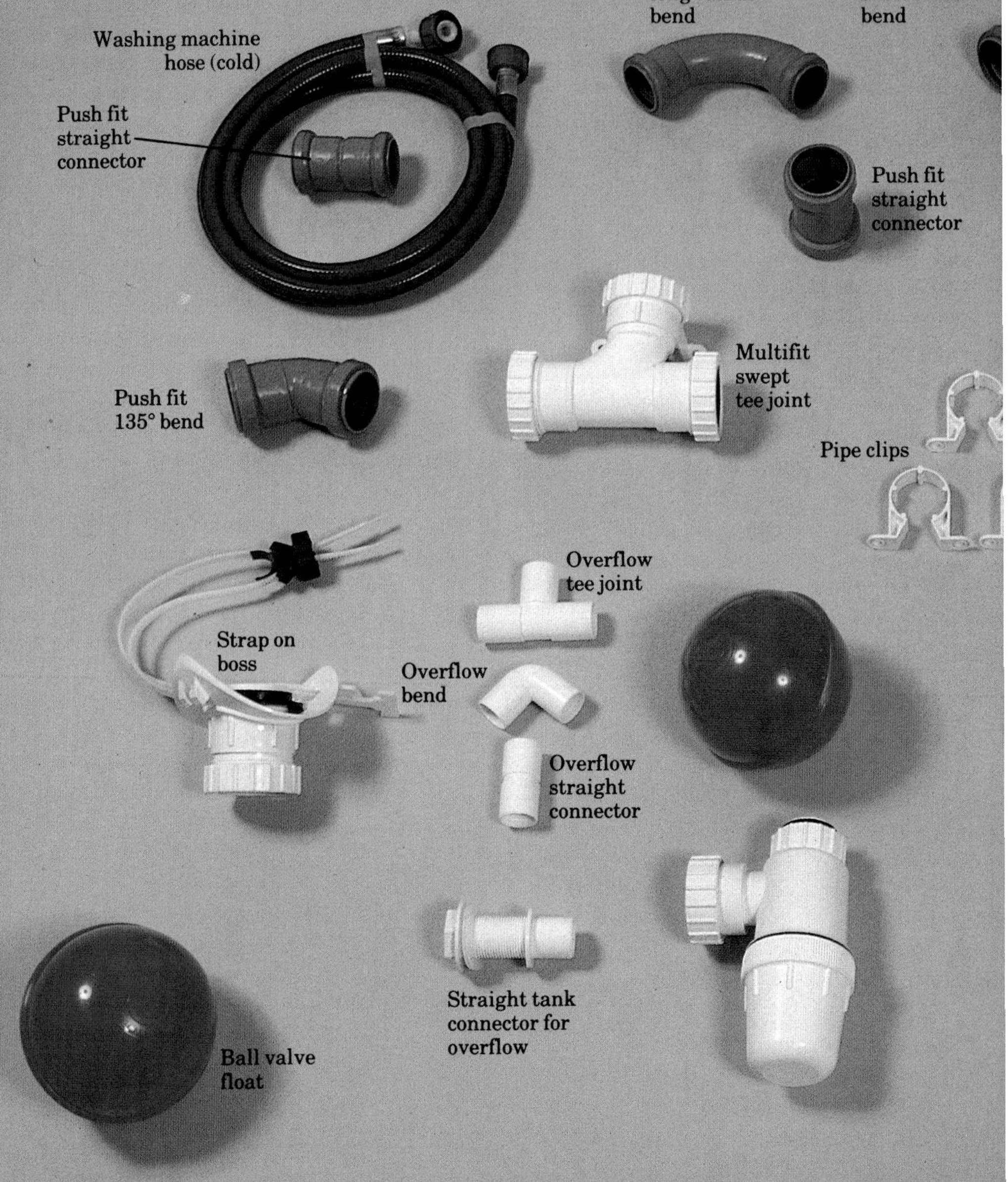

Washing machine hose (cold)

Push fit straight connector

Push fit long radius bend

Push fit short radius bend

Push fit straight connector

Push fit 135° bend

Multifit swept tee joint

Pipe clips

Strap on boss

Overflow tee joint

Overflow bend

Overflow straight connector

Straight tank connector for overflow

Ball valve float

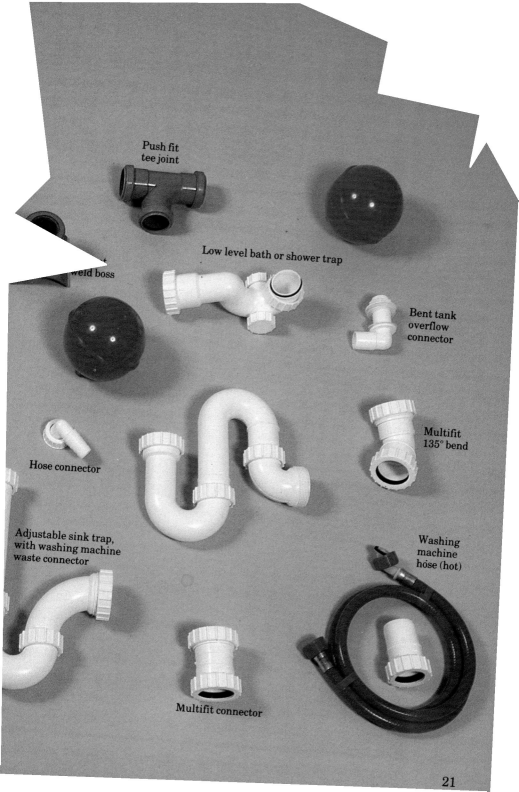

Push fit
tee joint

Low level bath or shower trap

weld boss

Bent tank
overflow
connector

Hose connector

Multifit
135° bend

Adjustable sink trap,
with washing machine
waste connector

Washing
machine
hose (hot)

Multifit connector

21

Repairs

An efficient trouble-free plumbing system is not merely a matter of luck. Repairs and maintenance are needed from time to time on all working parts.
This chapter describes a few of the jobs you can carry out to remedy the most common faults. Many repairs need only basic tools that you may already have in your tool box.

BALL VALVES
The problem of an overflowing tank or cistern is experienced by most people at some time or other and apart from wasting water it could cause damage through damp. Yet servicing a ball valve is a quick and easy job requiring few tools.

The first thing to do is turn off the water at the supply valve. If your valve is an old fashioned type that doesn't come apart like the ones shown here, it is probably at the end of its useful life and should be replaced. A mains fed valve should be replaced with a high pressure model and a tank fed valve with a low pressure model.

Undo the connector at the end of the tail where it joins the pipe. Remove the backnut that holds it on the tank or cistern and slide out the entire tail. When fitting a new one remember to replace the seals on the tank and the connector. If you haven't got a new fibre washer, wrap some PTFE tape around the connector end in place of the old washer.

Newer ball valves have a union nut which holds the body to the tail – undo this nut and take the body away. The piston type has a split pin holding the arm in position; this must be removed allowing the arm to come free. Take off the end cap and slide

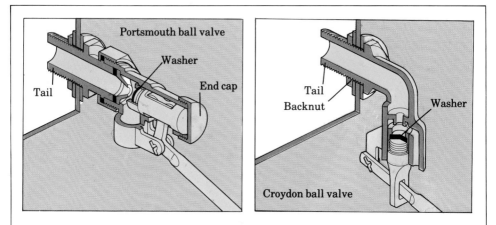

Most older ball valves are of the 'Portsmouth' or 'Croydon' type. It is a

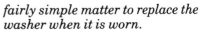

fairly simple matter to replace the washer when it is worn.

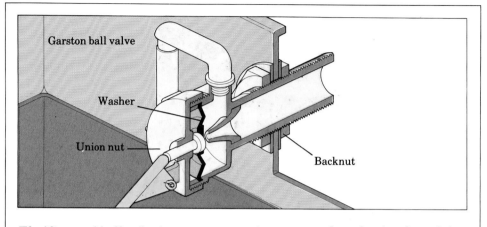

Garston ball valve

Washer

Union nut

Backnut

The 'Garston' ball valve is a newer type. It has a large rubber washer that is easy to replace, by simply undoing the union nut.

out the piston from its housing. Inside the piston you'll see a small rubber washer. To remove this unscrew the end of the piston using pliers or grips. It's not always easy to see that this section unscrews but it shouldn't be difficult to do. Once the end is free you can dig out the old washer with a screwdriver and replace it with one of the same size. You can make an emergency repair by turning the old washer round until you have time to buy a new one.

Before refitting the valve check the seating inside the body – if it is pitted the new washer won't work. Push out the old seating and fit a new one.

Newer type Garston ball valves have a large flat rubber washer which is easier to replace – simply undo the union nut and remove the worn-out washer. Make sure you fit the new one the same way round.

Finally, reassemble the valve and check it works correctly by turning on

the water – lift the arm to see that the water stops running.

If the ball float is damaged or has water inside it replace it with a new plastic float of the same size.

TAPS

Curing a dripping tap is often just a case of changing the washer. First turn off the water and check it has stopped running by opening the tap. Place the plug in to avoid losing the small screws down the waste hole.

Older-style taps have a shroud which must be loosened to get at the head. Sometimes you can undo this enough to get a spanner onto the nut underneath. If you can't, remove the tap top and slip off the shroud.

If the tap hasn't been dismantled for a while it may require a little extra effort. Hold the tap steady to stop it turning while you exert pressure. On some taps, a length of wood can be used as a counter lever.

If you have difficulty removing the gland on an old tap, try boiling water.

Use a wooden batten to brace the tap, and allow you to apply more pressure.

With the head removed you can get at the rubber washer on the jumper plate. Some are held in place by a small nut, others simply push on.

Replace the washer with the correct size. Don't be guided by the old washer since it may have spread through years of use. Sink and basin taps usually take a ½ in. washer although some are ⅜ in. Be guided by the jumper plate – the washer will be the same size. Bath taps take a ¾ in washer. Before replacing the tap check the seating inside the tap is clean and unpitted – a torch will help you examine this and your little finger can often feel any unevenness. If the seating is damaged you can fit a small plastic insert in most taps which often does the trick.

Supataps are a special design that can be rewashered without turning off the water; the lower half of the tap unscrews by turning anti-clockwise whilst holding the locking nut steady.

The jumper and integral washer is located in the end of the tap and is pushed out through the larger end of the nozzle. Fit a new washer into the anti-splash insert and replace the whole assembly in one piece.

Different size washers are avail-

Remove the tap handle or shroud to allow access to the gland unit.

Brace the tap and unscrew the gland nut to release the stem assembly.

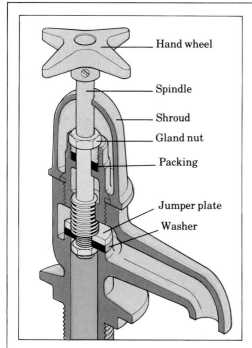

- Hand wheel
- Spindle
- Shroud
- Gland nut
- Packing
- Jumper plate
- Washer

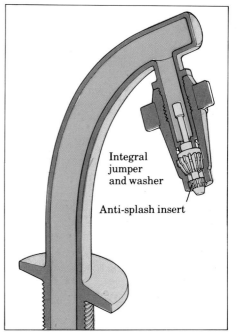

Integral jumper and washer

Anti-splash insert

Although taps come in many shapes and sizes, their internal design is

usually very similar. The exception to this is the supatap.

There may be a small screw or nut holding the washer in place.

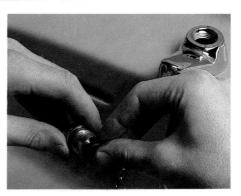

Having removed the old washer replace it with one of the same size.

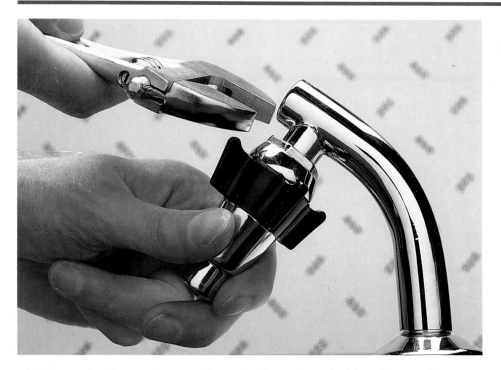

able to suit the various styles of supatap – it's essential to buy the right type, so you'll need to take the old one with you for comparison.

Apart from looking elegant, the supatap has easy washer replacement, without shutting of the water.

STOPCOCKS

Stopcocks also have washers and although you might not notice it the washer may have worn. This will become evident when you try to shut off the water and find the stopcock is still letting the water by. Whilst there is nothing difficult about rewashering a stopcock – the procedure is identical to rewashering a tap – the problem comes when you want to turn off the water. If the stopcock is the first one in the house you'll have to shut off the water underground.

The place to do this is at the water

board stopcock outside. Since this stopcock is the property of the local water authority you'll have to ask them to shut it off to coincide with your repair to the internal stopcock.

A common problem with stopcocks and taps is a leaking spindle. This occurs when the packing is worn and water seeps through the gland nut and down the spindle. Often it will happen when a stopcock hasn't been used for a long time. If you turn the stopcock off the leaking should stop. You can then loosen the nut and slide it up the spindle towards the head.

Repacking the spindle can be done

with a variety of materials – PTFE tape drawn into a string, knitting wool or string. Take a couple of inches of the packing material and wrap it around the spindle pushing it down into the housing. A small electrical screwdriver is ideal for doing this. Replace the nut, turn on the stopcock and test.

ADDING A STOPCOCK

To gain more control of your system it is sometimes a good idea to fit an extra stopcock close to a particular

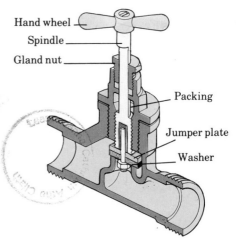

The internal design of a stopcock is very similar to that of an ordinary tap. Leakage or letting can be caused by a worn washer or a gland in need of repacking.

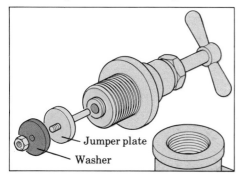

Common problems with stopcocks are either worn washers or gland packing.

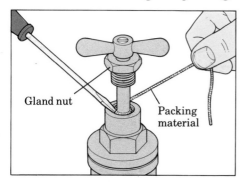

If washer replacement doesn't solve the problem, repack the gland.

fitment. This will enable you to carry out servicing without shutting off the rest of the house. An extra stopcock fitted near a ballvalve is an example of this – stopcocks made with integral compression fittings make it a very quick and easy job.

After turning off the water and draining the pipe through the lowest tap, mark out the length of the stopcock allowing for the pipe to enter at both ends. Cut the pipe with a hacksaw or cutter and slip on the compression fittings either end.

You'll notice a small arrow stamped on the stopcock body; this indicates the direction of flow – water will not flow backwards through a stopcock. Make the joints as described in the basics for compression fittings and test by turning on the water.

A new immersion heater

If your immersion heater suddenly stops working the chances are that it's burnt out. But before you rush out and buy a new one it's worth checking the wiring and the thermostat. The high termperatures inside the cover may have charred the flex at the terminals.

Before carrying out any work on the immersion heater turn off the electricity supply to the heater at the main fuse board. Remove the protective cover from the top of the heater to expose the wiring terminals. If the connections are secure and the wires are in good condition it is most likely that the heater itself is at fault. You can test the element inside by using a bulb circuit tester.

Place the contacts on the terminals, one on the live and one on the neutral; if the bulb lights up it means a circuit is being made through the element so it must be sound. This means the fault is in the wiring or the thermostat and you should test it.

Assuming the heater is faulty the cold supply to the cylinder must be turned off at the gate valve. Drain the pipe work through the kitchen hot tap. There will still be water in the cylinder and a little of this must be drained. At the bottom of the cylinder you should find a draincock on the cold feed in. Attach a garden hose to the spout and run it to a lower point – ideally to the garden. Undo the small square spindle on the drain cock until water begins to run – be careful not to undo the spindle more than a couple of turns as you may remove it altogether. Once a gallon or so has drained from the cylinder you can safely remove the heater, (unless it is positioned in the side, in which case you must drain the whole cylinder).

Removing the heater requires a special spanner which can be bought fairly cheaply or hired from a good hire shop. Disconnect the wires to allow you to undo the heater; an old heater may be difficult to unscrew. Don't be tempted to apply extra lever-

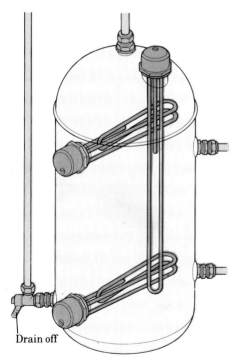

Drain off

A cylinder may have one immersion heater running vertically, or two horizontal ones to heat different amounts of water.

age to the spanner as you may distort the cylinder boss. Gentle heat from a blowlamp around the immersion boss will help ease the threads. The copper will expand slightly and you can then apply the spanner and try again – use a piece of rag to protect your hand.

Once you've removed the heater measure its length so you can buy an exact replacement. Don't forget to buy a thermostat at the same time unless you're planning to use the old one.

A sealing ring is supplied with the new heater. There is no need for any sealing compound to be added; so long as the cylinder boss is clean a good seal can be made. Save the last quarter turn until you have filled the cylinder and tested with the hot taps closed – it's better to tighten down until the seepage stops rather than overtighten before the water is in. Secure the wires onto the new terminals, remembering to thread the grommet over the flex. The most important wire is the earth – make sure this is wrapped around the terminal and twisted to prevent it coming loose. The neutral fixes directly to the heater but the live must go to the thermostat, a short wire on the heater goes into the other side of the thermostat to complete the switched circuit. Set the temperature on top of the thermostat to around 60° C before fixing the protective cover.

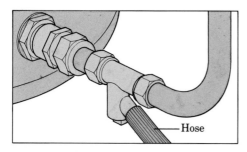

Attach a hose to the drain cock, to lower the water level in the cylinder.

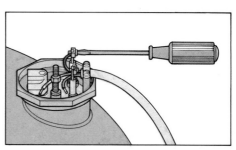

Remove the cap and disconnect the wires from the terminals.

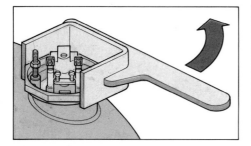

You may need to apply gentle heat, before the element will unscrew.

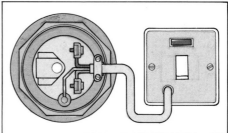

When the new element is in place, reconnect the wires to the terminals.

Changing taps

Changing your old worn out taps for one of the exciting new styles is a good way to test your plumbing skills and give you confidence to carry out more ambitious projects in the future. Before buying the taps bear in mind that bath mixer taps should not be connected to the rising main as there is a risk of cross flow from stored water. If you wish to use mixers on a bath the cold supply must be fed from the storage tank.

An elegant selection of modern taps for baths, basins and sinks. This range of *taps all have gleaming chromium plated heads.*

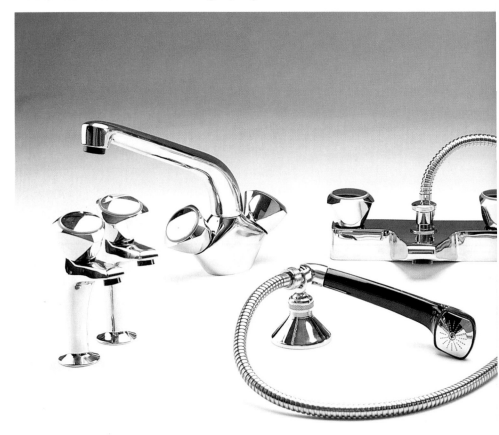

Below: Shower mixer and shower kit with flexible chrome hose.
Below centre: Sink mixer taps are also available in colours to match your kitchen decor.
Bottom: This ultra-modern bath mixer unit is finished in white.

Bidet mixer with adjustable nozzle and pop-up waste.

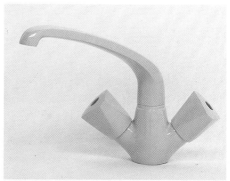

31

Changing taps

Start by turning off both the hot and cold supplies and opening all the taps in the house to drain water from the pipes. A special basin spanner is required to undo the connectors on the old taps.

Because space is limited you'll need to find a comfortable position to work in. Try lying on your back with a cushion under your head. A light of some sort will help you see the nuts. Using a lever in the spanner loosen the nuts that connect the taps to the pipework – a little water may come out which you can catch with an old rag. The back nuts on the taps may be quite stiff; if they prove difficult try some penetrating oil on the threads. Ask someone to hold the taps steady whilst you undo the backnuts.

All taps are provided with a soft washer of some kind to go between the tap and the surface of the sink, basin or bath. Mixer taps have a one piece rubber seal which must be fitted before the taps are positioned. Sink mixers also need a top hat washer on the underside of each tail to act as a spacer. The backnuts can then be fitted and tightened.

Brace the tap with a batten before removing the nut from the shank.

You need a basin spanner to reach the recessed nut on the tap shank.

An adjustable wrench can be used to loosen the nut on the tap connector.

When the tap is removed, clean off any sealing compound left behind.

It's likely that you'll find your new taps are slightly shorter than the old ones leaving a gap between the connectors and the tails. Sometimes there is enough play in the pipework to overcome this but don't be tempted to strain the pipes to meet. A simple solution is to fit a pair of tap extenders made especially for the job. These screw on the tails with a fibre sealing washer placed in between the tap and extender. To make doubly sure of a good seal wrap a few turns of PTFE tape around the threads before screwing on the extenders.

With the extenders in place the tap connectors can be tightened on, but first check the fibre washers are intact at the top of the connectors. It's far easier to replace them with new ones now than risk a leak when the job is finished. If you don't have any fibre washers wrap a good quantity of PTFE tape around the threads.

Tighten the connector nuts on to the new taps or extenders – the job is now complete and ready for testing. If you find that the hot and cold are on the wrong sides don't worry, the taps are both the same so you can simply swap over the tops.

With the connector fitted to the tap shank, mark the pipe for cutting.

Cut the pipe to length, and clean the pipe end with abrasive paper.

Slide the nut, then the olive onto the pipe. The joint can then be assembled.

Having applied a few turns of PTFE tape, screw the connector to the shank.

Use two wrenches to tighten the nuts, 1½ turns should be enough.

Fitting a sink

A new sink can do a lot to revamp an old kitchen. By taking advantage of the latest styles and materials you can bring a fresh look into an area of your home that can otherwise easily begin to look dreary. Whether you intend to change just the sink or make it part of a more ambitious plan, the basic job is the same.

If you're fitting a new sit on sink to an old base unit check the width and depth since the new ones are metric and won't fit old imperial units.

When buying a new sink you'll also need a waste fitting. Some sinks have a larger hole which can be used for a waste disposal unit. If you don't require this a conversion fitting called a basket strainer waste will bring the waste down to a standard 1½ins. For standard hole sinks buy a combined waste overflow fitting.

It's easier to make the waste up before you fit the sink in place. The waste can be bedded on putty or non-toxic mastic to provide a seal between the metal surfaces. The overflow collar goes on the underside of the sink and is held in place by a backnut. Place a nylon washer and some mastic in between the backnut and the overflow fitting to prevent any leaks. You may need to stop the waste fitting turning while you tighten the backnut – this can be done using two screwdrivers cross sword fashion.

Before you fit the sink in place attach the taps, if you're using flexible tap connectors you can also fit them as well.

The sink can then be fitted to the base unit. Sit on types have a lip on the underside to take retaining clips attached to the base unit. When they are all in position the clips can be evenly tightened to hold the sink

firmly in position.

You may wish to let the upstand at the back of the sink into the wall, in which case you'll have to mark and chase out the wall with a chisel before the sink is finally fixed.

INSET SINKS

Inset sinks are a little more involved because they require a hole in the worktop. It's best to secure the worktop to the base unit before you begin cutting. This will hold the worktop steady and allow you to see exactly where the sink should go.

The most important part of this job is planning. Make a careful check of the underside of the worktop so you can avoid positioning the sink bowl over a drawer or structural support and take into account the position of the front rail so the sink is placed behind it.

When you're happy that the sink will fit, mark out the worktop with a pencil. Draw a couple of parallel lines on the top squared off with the front then take the sink or template and draw the outline on the worktop. A template gives you the cutting line but if you're working without one you'll need to draw a cutting line within the outline. Measure the distance between the edge of the sink and the fixing toggles – it will be something like half an inch. Getting it right now will save you some laborious filing later on. Once the worktop is marked out drill a pilot hole inside the cutting line. You can

This elegant and practical inset bowl and drainer are complemented by the matching white mixer unit.

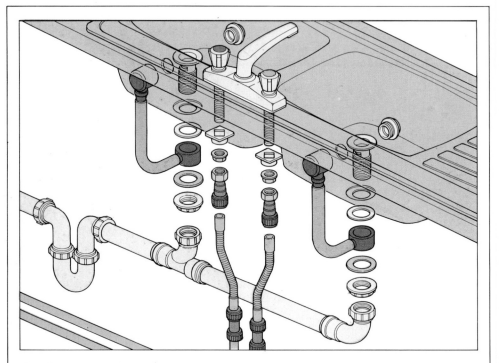

The illustration above shows a typical supply and drainage layout for a double sink. Some modern units are designed to accommodate a waste disposal unit, and others may have a separate draining tray.

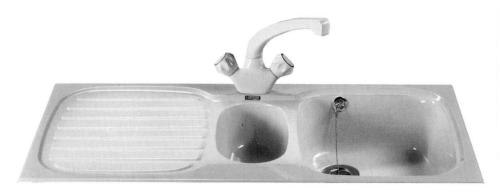

Just one of the wide range of sink units available. There are many variations of single and double bowls or drainers to suit all requirements.

cut the hole with an electric jigsaw or a padsaw, but using a padsaw is quite hard work.

Work carefully around the line; if you stray go back a bit and correct your course. Don't place your hand under the worktop, instead support it with a piece of wood to prevent the cut out breaking off. When the hole is cut try the sink in position and make any adjustments before fixing. When you've cut the hole, seal the open surface of the chipboard with polyurethane or oil based primer to prevent it taking up moisture and swelling.

All inset sinks need some kind of seal between the edge and the worktop. This will be provided with the sink either in the form of a mastic or neoprene rubber strip.

Making sure the sink is correctly seated, turn the toggles tightening them alternately – avoid distorting

Cutting the hole for an inset sink can be done with a padsaw or more easily with an electrically powered jig saw.

the sink by uneven tightening.

The last job is to attach the trap and waste pipe. Screw the trap on the threaded waste with a rubber washer inside to make the seal. Decide on the best route for the pipe and connect it to the trap – if it's to go downwards you may use a trap bend to change direction. On push fit traps apply a little washing up liquid to the end. See Basic Techniques on running waste pipes.

Inset sinks and drainers are neat and elegant for today's modern kitchen.

They take a little longer to fit, as they require a hole to be cut in the worktop.

A new bathroom suite

There are few home improvements you can make that will have a bigger impact than a gleaming new bathroom suite. The great range of colours and styles enhances the feeling of luxury you experience when lying back in a new bath. When the plumbing is finished you can add your own touches in the tiling and decor to create a completely new environment.

Whatever you decide to do, be it a straightforward change over of fitments or a more ambitious rearrangement, the key to success lies in careful measuring and planning to make sure you choose the right bathroom suite for your requirements.

At Texas you can browse around the display suites at your leisure, checking dimensions against a sketch of your bathroom, so you are satisfied everything will fit before you buy.

When you get your new bathroom suite home, make up all the fitments before you remove the old suite. This will involve fitting the supporting legs to the new bath and assembling the waste and overflow – you can even fit the taps ready for connection to the pipes. The more you can do at this stage, the more it will cut down the time you're left without facilities.

Bathroom suites need no longer be plain colours, as these two exciting examples demonstrate.

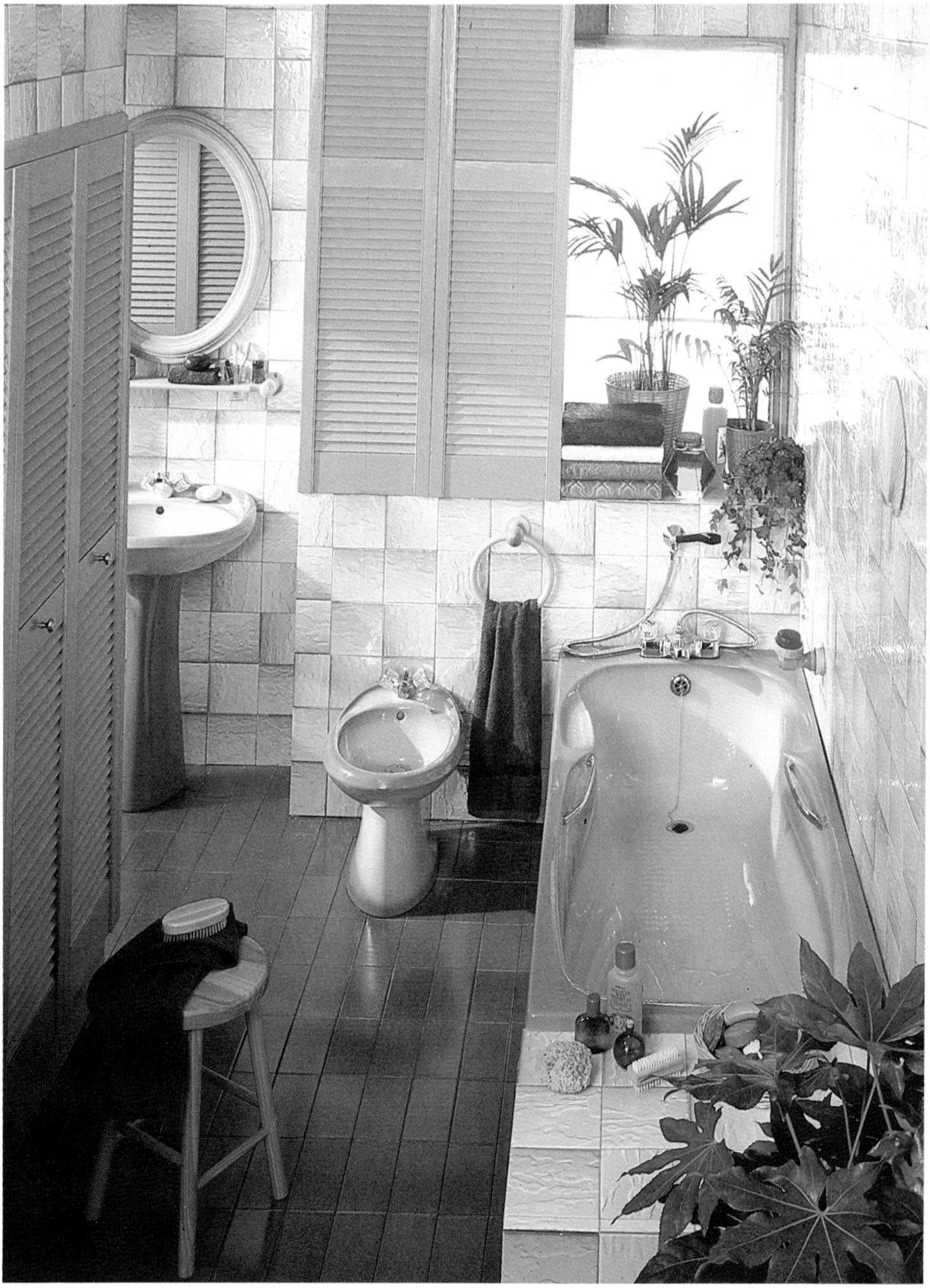

FITTING A NEW BATH

Most new baths are supplied with full instructions on how to fit the supporting legs. Once this is done you can fit the waste and overflow – take care to fit rubber sealing washers on the underside of the waste and overflow holes. If your waste doesn't incorporate an overflow buy a special bath trap and overflow assembly. Don't overtighten the centre screw that holds the waste together – it's better to give it a little extra turn when the bath is tested if you find it seeping.

REMOVING AN OLD BATH

Removing the old bath can be done very quickly if you simply cut through the old connections with a junior hacksaw – remember to turn the water off and drain the taps before you start.

If the old bath is made of cast iron it will be very heavy; rather than struggle down the stairs risking damage to the decorations simply break it up with a heavy brick hammer – wear protective goggles and a heavy pair of gloves. You may also want to protect your ears with cotton wool – the bath will resound like a bell when you strike it. Work from the edge breaking small pieces off at a time and carrying them outside as you go and, finally, sweep up the porcelain enamel splinters so the site is safe and clear ready for the new bath to be plumbed in.

Older houses will almost certainly have lead tails that connect the iron pipe to the taps – these are best discarded at this stage. You may also wish to replace the iron piping with copper. Fitting copper to galvanized

This Bahama bath in Alpine Blue adds a touch of luxury to the bathroom.

iron will set up an electrolytic reaction accelerating corrosion in the iron so they will eventually need replacing anyway.

Fit flexible (22mm) tap connectors to the bath taps and arrange the rigid pipework to terminate close to the connector ends. See the note in Basic Techniques about converting old ¾in pipes to 22mm.

Use compression rather than capillary fittings – a blowlamp may damage the bath. If your cold supply is mains fed it may be 15mm (1½in) – a 15mm to 22mm compression coupler will make the conversion to the larger size of pipe.

Level the bath with the adjustable

feet before screwing the wall brackets on – check the ends and side with a spirit level. If the bath doesn't fit evenly against the wall chase a small channel in the plaster and let the bath in slightly, but if you're planning to retile the bathroom the tiles will probably cover the gap so it's not important.

Once the bath is in position you can fit the waste pipe and trap. If the old pipe is lead it's wise to replace it with a plastic system which is simple and quick to install. Connecting up the waste outlet isn't difficult since the waste usually feeds into an external hopper.

Remember to make good the hole in the wall around the pipe – this will stop an icy draught freezing the water in the bath trap, and also help to reduce your heating bills.

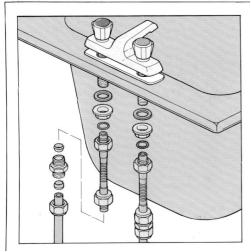

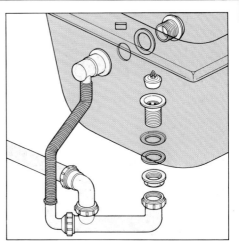

This illustration shows supply and drainage connections for a bath. Plastic drainage fittings and flexible

tap connectors make fitting a new bath much simpler. Always ensure that plastic fittings are not overtightened.

FITTING A NEW BASIN

The most popular choice for a basin is the pedestal type; this has the advantage of hiding pipework and overcoming the problem of fixing brackets to the wall.

Make up the basin with waste and trap before fitting it in position. If there's a pedestal retaining bracket it should be fitted before the waste backnut. Bed the waste on mastic or putty. Use a 1¾in slotted waste to take the integral overflow. Fit a nylon

A Bahama pedestal basin in Avocado. The pedestal neatly conceals the pipework and the waste trap.

washer under the basin before you screw on the backnut. You may have to stop the waste rotating in the hole by inserting two crossed screwdrivers through the plug hole. You'll need a fairly large spanner or pair of grips to tighten the nut but avoid excessive pressure which may squeeze out all the sealant and even crack the basin.

Attach flexible tap connectors to the fitted taps and offer the entire assembly of basin and pedestal up to the chosen position against the wall. If you can get someone to steady the basin while you make adjustments you'll find it a lot easier. A small amount of non-setting mastic around

the top of the pedestal will help keep things in position while you're marking the fixings.

It's important to make the basin level across the back where it meets the wall. Because of the shape of some basins you may not find it easy to sit a spirit level across the back, in which case try resting it across the two tap heads.

Draw a pencil line around the back of the basin and mark the two screw holes on the underside. Holding the

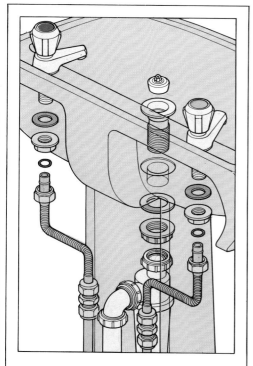

Flexible tap connectors and a bottle trap fit inside the basin pedestal, for a neat and unobtrusive finish to your bathroom.

basin in position mark the pedestal outline on the floor. If the pedestal doesn't appear to sit correctly, it's probably because the floor isn't level. You can overcome this by packing up one side of the pedestal with pieces of old floor tile or similar. Once the basin is finally fixed the packing can be replaced with filler.

Check the basin from the front to make sure the flexible connectors are bent so the basin and pedestal conceal them, then mark the pipework for cutting. If you haven't yet run the pipework mark out the approximate positions giving height from the floor and distance from the wall. You can then take the basin and pedestal away while you assemble the pipe runs.

Drill and plug the wall to take two 2in screws and mark the holes on the floor. If it's a solid floor you will have to drill and plug it.

Connect up the supplies with compression fittings and fit the waste pipe into the trap, before testing the installation. When you're sure that the basin won't have to be removed again you can apply a silicone seal along the back edge. If you're tiling a splash back you can save awkward tile cutting to fit around the basin by placing them behind the basin and sealing the joint afterwards.

Inset basins for vanity units are similar to inset sinks (see page 37). The basin is held in position with a mastic strip or silicone sealant applied before bedding the basin on the top of the unit. Ensure the hole is large enough to allow the basin to sit squarely before you begin to apply the sealant.

This Trinidad close-coupled WC has graceful sculptured lines and is available in a range of colours.

FITTING A NEW WC

Fitting a new WC is a job that requires some disruption of essential facilities. If it's the only WC in the house you'll want to carry out the job quickly. This will require careful measuring and planning to make sure everything will fit before you remove the old suite.

New WC pans have horizontal outlets which will have to be converted to suit the individual installation. This is done by using a pan connector which pushes into the existing soil pipe making a 'P'trap out through the wall or an 'S'trap going down through the floor.

If you're fitting a close-coupled suite where the cistern sits on top of the pan there's less scope for adjust-ment of the pan position. Instead you'll have to alter the soil pipe either extending it by fitting a straight push fit extender or shortening it by cutting the cast iron pipe with a hacksaw. If you're not prepared to do this the cistern can be packed out from the wall on spacer blocks or timber battens but the final job won't be as neat.

Make up the cistern with the flushing mechanism and handle on the same side as the ballvalve connection. On many models this can be done on either side to suit existing pipework. Be sure to fit the sealing washers provided, on the inside of the cistern – there's no need to use mastic or additional sealant.

With the pan in position offer up the cistern and mark out the overflow and supply pipes. In the majority of cases the overflow will go directly out through the wall with a slight fall. If you're not lucky enough to use the old overflow position you'll have to drill a new hole. Check outside for any obstruction such as the soil pipe before you begin drilling. You'll need a hole $7/8$in minimum diameter to push the plastic overflow through.

Connecting the cistern to the pan is done in one of two ways. The flush pipe is pushed into the back of the pan with a rubber cone seal – a drop of washing up liquid will help – and the top of the flush pipe enters the underside of the cistern and is sealed like a waste pipe with a nut and rubber ring. This need be hand tight only, grips may distort the plastic.

Close-coupled cisterns bolt directly on to the pan with a large rubber washer sandwiched between the pan and cistern. Tighten the bolts evenly

The cistern can be placed at any level with a standard washdown WC. The

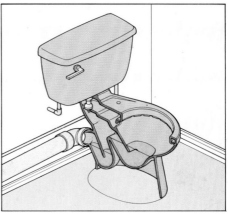

close-coupled cistern is bolted directly onto the pan.

so the cistern sits level. When it's correctly positioned mark the wall fixings then remove the cistern while you drill and plug the holes. Fix the cistern to the wall using brass screws. A couple of ¾in tap washers will make good protective washers under the screw heads.

The supply pipe is joined to the ballvalve tail by a tap connector. It's best to use a compression fitting to avoid damage from a blowlamp. Apart from the small fibre washer in the tap connector no other seal is required to connect up the ball valve. As most WC ball valves are now made of plastic take care when tightening the nut to avoid damaging the thread.

An alternative plastic insert is supplied to convert the ballvalve to low pressure. If your supply is mains fed leave the small hole insert in place. If it's fed from the tank simply change over the insert.

A push-fit WC waste connector can be lubricated with silicone grease for ease of fitting.

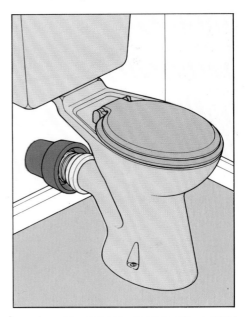

Adding a shower

Besides being quick and refreshing a shower is one of the best energy saving investments you can make. For the cost of taking a single bath you can have up to six good showers. Quite apart from saving money there is the convenience of stepping into a quick shower. By fitting one in a bedroom for example, you can bypass the morning traffic jam at the bathroom door. But before you buy one think about which type will suit you best.

Basically, there are two types of shower – the electric instantaneous sort and the kind that you link up to your existing hot and cold plumbing.

Electric showers are fed from the mains water supply so there are no problems with the height of your storage tank – so long as you have a good mains pressure an electric shower will operate satisfactorily. The water is heated as it passes through the unit which means no waiting for centrally heated water to heat up. The amount of power consumed by an electric shower is quite high but only for a very short period, nevertheless you'll require a hefty power cable of 6mm^2 minimum taken directly from the fuse board. If you're not happy about this part of the job you can complete the plumbing work and ask an electrician to carry out the wiring.

When you've decided upon the location of your shower think about the easiest route to run the cold supply; very often the best way is to tee off the rising main in the loft before the ballvalve. Turn off the water and drain the pipework by opening the cold tap in the kitchen. Press the ballvalve down to release the water held in the pipe. Cut into the pipe and fit a 15mm equal compression tee. It's advisable to fit an isolation valve just after the tee which will shut off the

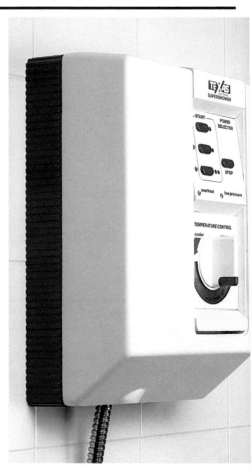

The Texas Supershower offers the luxury of instantaneous hot water for safe and efficient family use. It also saves money on your heating bills.

supply for maintenance.

Most shower manufacturers recommend a gate or Ballofix type valve rather than a stopcock with a loose jumper. This is because an electric shower shuts off abruptly on a solenoid switch and it can cause noises in the pipework. It's essential that the pipework is well clipped to prevent knocking noises. Run the 15mm pipe to a point over the shower unit. Poke a small screwdriver, or something similar, through the ceiling to indicate the position of the pipe. The supply pipe can be surface run neatly down beside the shower unit or in plastic trunking, which may also house the supply cable.

The shower unit itself should be placed out of the direct spray but still within the enclosure so you can conveniently adjust the controls. Make the connection to the pipework with a compression fitting which is usually supplied with the shower.

Turn on the water supply and test the pipework for soundness, then insulate anything exposed to outside temperatures.

The electricity supply must come from an independent 30 amp fuseway, ideally protected by an earth leakage circuit breaker. The shower instructions will give the exact size of the cable depending upon the kW rating of the unit. A pull cord double pole ceiling switch with a neon indicator should be fitted to enable you to switch off the shower after use.

Before using the shower check that an earth clamp is fitted to the shower pipework and a single 6mm^2 earth cable is run back to the house's main earth terminal at the fuse board.

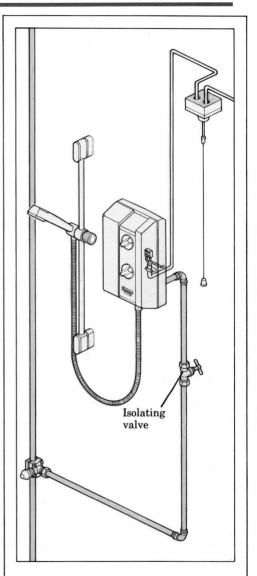

Isolating valve

Fitting an electric shower is not a difficult job for the DIY enthusiast, providing that the manufacturer's instructions supplied with the unit are followed carefully.

PLUMBED IN SHOWERS

The other type of shower is fed from a cold water tank and a conventional hot cylinder. It's important that the hot and cold are both at equal pressure; this is assured if the cylinder is fed by the same tank that is supplying the cold feed to the shower. The cheapest mixer is a straightforward manual type; this is an unsophisticated but reliable device, but it will fluctuate in temperature when other plumbing is used at the same time. To overcome the worst effect of this i.e. being scalded, you must fit an independent cold supply straight from the tank. This is easy on plastic tanks where a hole can be readily made for a new tank connector (see Fitting a new tank). On older galvanized iron tanks it's rarely worth the trouble of adding a new connector; instead choose a thermostatic shower mixer.

These models have the ability to compensate for temperature fluctuations giving you a safe and comfortable shower which is virtually unaffected by people using plumbing elsewhere. They are a must for children or old people and will often repay the extra cost in simplified plumbing.

All you need to do is find the most convenient point for teeing off the hot and cold supplies to your bath. Where the cold tank is less than one metre above the shower head it's advisable to run the supplies to the shower in 22mm and reduce down to 15mm for the short length leading to the connections only. This will help overcome the resistance in the pipework so you'll still get a reasonable flow. The best solution to poor pressure in mixer showers is to raise the cold water

The Silver Spray shower set has a flexible hose and sliding rail. It is available in a range of colours.

tank on a sturdy timber platform.

Inside the shower area the pipes may be run in chases and be plastered over ready for tiling. The chases can be cut by hammer and chisel or by using a power tool available from hire shops. Wherever possible avoid joints in pipes to be buried – test them by capping off and filling. Coat the pipes with bitumastic paint or wrap Sylglas tape around them to protect the copper from corrosive acids in the cement and plaster.

The pipes can be left capped and protruding through the wall in position for the shower mixer when tiling of the walls is complete. Take care not to drill through buried pipes when fitting the mixer and spray head.

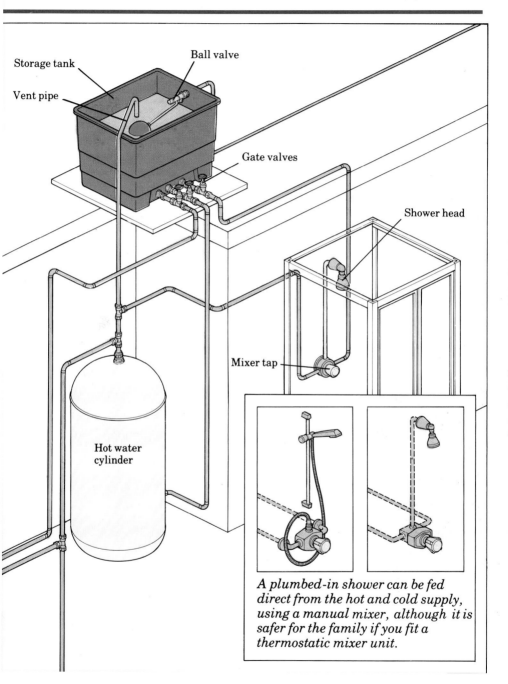

Storage tank

Ball valve

Vent pipe

Gate valves

Shower head

Mixer tap

Hot water cylinder

A plumbed-in shower can be fed direct from the hot and cold supply, using a manual mixer, although it is safer for the family if you fit a thermostatic mixer unit.

FITTING A SHOWER TRAY

There are several ways of fitting a waste pipe to a shower tray; a lot depends upon how much access you can prepare. If the shower tray is going onto floorboards it's often possible to cut away the board for the trap and then remove another board next to the tray. This will enable you to stretch an arm under the boards and fit the waste pipe to the trap. Make sure you know what you have to do because you'll be working blind.

Where the waste pipe goes directly through a wall you can cut a hole in the side of the tray nearest the wall so the waste pipe can be fitted to the tray before it is slid into position. The waste pipe must be long enough to reach all the way through the wall. Fit an access tee so you can clear the

Shower cubicles are made in panels, and can be supplied to suit any situation and layout. Shower trays have no overflow, so make sure that the trap is accessible, as a blockage could have serious results.

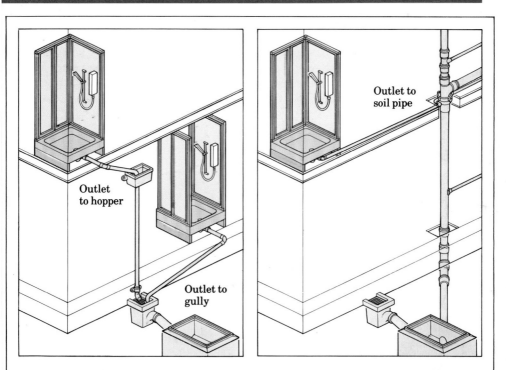

Depending on the layout, shower waste pipes can be connected direct to the soil pipe or via a hopper. If your shower is located on the ground floor the waste can be discharged into a nearby gully.

trap without removing the tray.

On solid floors you may have to build a small plinth of 2in x 2in timber so the trap clears the floor. By making one side removable you'll always have good access to the trap – it's essential that you have access to traps because hair deposits can block a trap in as little as six months and as showers have no overflow the build up of water in the base will soon cause a problem.

A variety of brackets are supplied with the various types of shower tray to secure it to the floor. If these prove impractical in your situation bed the tray on a bead of silicone sealant or non setting mastic. The most important thing is to get it level. If the floor is uneven this may mean packing one side up – be sure to maintain even support of the base to prevent movement and possible cracking.

To finish off your shower a great variety of screens and doors is available from Texas complete with fitting instructions. Some of these are even pre-plumbed complete with the shower fitment so all you have to do is connect up the supply and waste pipes, and it's ready to use.

Fitting a cold water tank

An old and rusty steel tank is bound to need replacing sooner or later – carrying out the job when it is convenient to you is far better than leaving it until it becomes an emergency. New plastic tanks are lightweight and easy to work with. Fitted correctly they are made to give a lifetime's service.

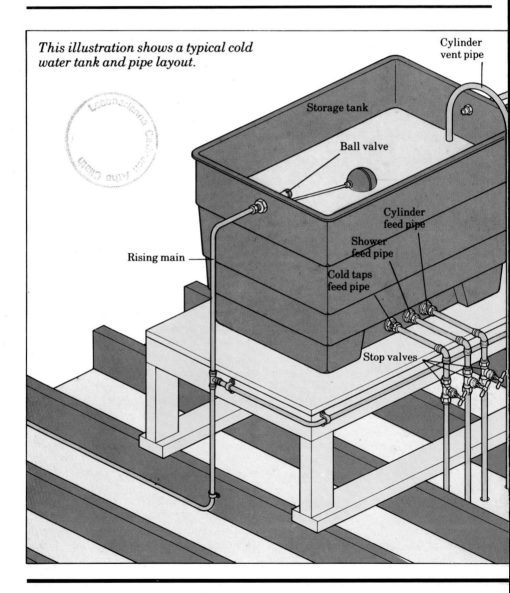

This illustration shows a typical cold water tank and pipe layout.

Cylinder vent pipe

Storage tank

Ball valve

Cylinder feed pipe

Shower feed pipe

Rising main

Cold taps feed pipe

Stop valves

Removing the existing tank is simply a matter of turning the mains water off and draining down the tank sponging out the residue into a bucket. The pipes can be cut or undone, whichever is easier.

Cold water tanks are sold in diffe-

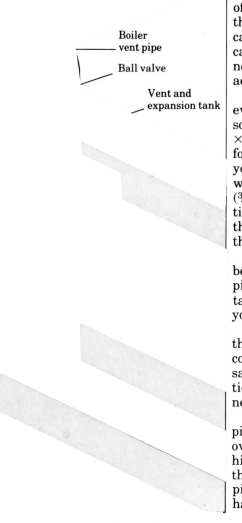

Boiler
vent pipe

Ball valve

Vent and
expansion tank

rent sizes to suit varying needs. If your cold water tank does nothing more than supply the hot water cylinder a 25 gallon tank will usually be adequate. Where there is also a cold feed to the bath, basin and WC you should select a minimum size of 50 gallons.

Before buying a tank check the size of your loft hatch and see what will go through. Tanks are described in two capacities – the nominal size is the capacity to the brim which it will never achieve, the smaller size is the actual capacity for working purposes.

Water weighs 10lbs per gallon so even a small 25 gallon tank will need some support. Use two 100 × 50mm (4 × 2in) timbers spread across at least four joists to distribute the weight. If you can straddle an internal wall this will help greatly. Lay a piece of 19mm (¾in) ply or blockboard across the timber bearers to support the base of the tank making sure you screw it to the bearers.

Make up all the fittings on the tank before you begin connecting to the pipework. This can be done before you take the tank into the loft but only if you're sure they won't foul the hatch.

The positioning of connections to the tank is an important factor in its correct operation. To save unnecessary pipe runs try to keep the connections on the side nearest the pipe you need to join.

To avoid having to alter the vent pipe, position the tank so it hooks over the side. If your new tank is a lot higher or you wish to raise the tank then you will have to alter the vent pipe anyway. Usually this means you have to add a middle section

Fitting a cold water tank

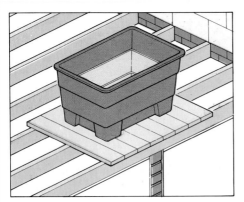

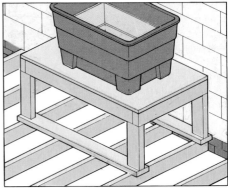

A 50 gallon tank weighs 500lbs when full. Ensure that it is above a load- *bearing wall or build a platform, spanning at least four joists.*

and reusing the bend in a new position.

The outgoing tank connectors should be fitted about 2in above the base of the tank to prevent sediment being drawn into the pipes – check to see that you don't drill through an internal bracing support. Make the holes using a hole saw slightly larger than the tank connector. Support the wall of the tank with a block of wood, clean any swarf and thread through the connectors. Fit a polythene washer either side of the connectors to prevent the brass contacting the wall of the tank. As an extra seal wrap a few turns of PTFE tape around each connector close to the polywasher. Tighten on the brass backnuts so they are firm but not so tight that they distort the walls of the tank.

The ballvalve can be fitted to the top edge of the tank with at least one clear inch above the hole. Fit the ballvalve with polywashers in the same way as the tank connectors. Before fitting the backnut place the

bracing plate supplied with the tank over the ballvalve tail. This will help to prevent excessive movement of the valve when the tank is filling. The pipework can be connected to the valve by means of a ½in tap connector and fibre washer.

It's advisable to fit a stopcock to the pipework immediately before the valve to give an independent means of shutting off the supply to the tank.

The plastic overflow connection must be at least 1in below the ballvalve inlet. The overflow connector is fitted in the same way as the other tank connectors and must be made equally watertight. Run the overflow pipe out through the eaves where it can be easily seen if it begins to discharge excess water. You should be able to thread it through the same hole as the old one. Support the overflow pipe to prevent it sagging if it fills with water. A simple system of pipe clips hanging from timber battens is an ideal way of supporting an overflow pipe. The battens can be tacked to the side of a rafter.

The final connections to the tank are the cold feeds to the cylinder and taps. The 22mm or larger pipes are joined to the tank connectors by compression nuts and rings. Support the pipes soon after the fittings to prevent any strain on the tank walls. It's a good idea to fit gatevalves fairly close to the tank so that the complete run of pipe work can be isolated.

Before you insulate the tank and pipes test the installation and check for leaks. Adjust the ballvalve so the water line is about 100mm (4in) below the top of the tank. Older style ballvalves are adjusted by bending

If your loft hatch is fairly small you may find that a flexible, round tank is easier to get through.

the float arm until the water stops. New types have an adjustment screw on the float to move it up and down on the arm. When you're happy with the operation of the ballvalve test the overflow pipe by pushing the float below the water line. The overflow pipe must be able to cope with the excess water. If the tank fills too rapidly turn down the supply stopcock to reduce the flow.

To prevent pollution of the water a lid must be fitted to the tank which can be bought when you buy the tank.

Mark out the position of the vent pipe on the lid and cut a hole so the pipe can be manipulated through. Finally, wrap all the pipework and the tank with insulation – remove any insulation from the underside of the tank so that it benefits from the heat in the house.

Adding a garden tap

An outside tap is useful for many jobs from car washing and
concreting to watering the garden.
If you've ever tried using a hosepipe from the kitchen tap you'll
appreciate how a garden tap can make life easier all round.
Fitting one is a quick job and even if you have to hire a drill,
it still works out surprisingly inexpensive.

An outside tap can go practically any-
where you like but the most obvious
choice is above the drain gully that
takes the kitchen waste pipe, since this
is also likely to be near the cold supply
to the kitchen sink.

When you've located the rising
main inside the house, work out the
shortest route for the new pipework.

You'll need to measure the position
for the tap from a reference point that
can be seen from both sides of the
wall. This could be the sink waste
pipe or the kitchen window – check

*An outside tap is very convenient and
can provide a trouble-free hose
connection, with a screw fitting.*

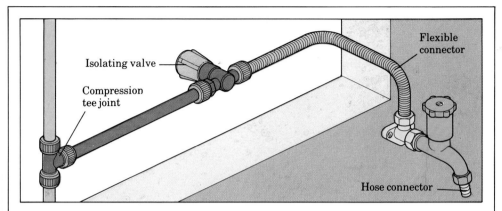

Isolating valve

Compression
tee joint

Flexible
connector

Hose connector

A position above the drain gully is ideal for an outside tap, with easy access to the water supply in the kitchen. A stop valve is useful to allow you to isolate and drain the tap in cold weather to prevent freezing.

that the waste pipe doesn't obstruct the hole you'll have to drill. You can hire a 20mm drill at least 300mm long to make sure you go through in one go. Avoid any rise on the pipe as it goes through the wall which may prevent the tap being drained in the winter.

The tap – known as a "hose union bib cock"- is screwed into a backplate elbow. It's easiest to fix the elbow to the wall before connecting the tap.

To minimize the exposed pipework arrange the backplate so you have room for one male/female copper elbow going from the backplate into the hole. Join the elbow to the pipe and slide the complete assembly into place. If you have no blowlamp and don't wish to purchase one for a small job you can prefabricate the pipe run by soldering the elbow on over a gas cooker ring.

The male fitting should slide into the backplate elbow and be secured by the compression nut. Mark out the holes for the backplate and swivel the fitting aside while you drill and plug them. Use brass screws to secure the backplate to the wall.

The tap can then be fitted, wrap about five turns of PTFE tape around the thread and screw the tap into the wall plate.

Inside the house continue the pipe to the point where it joins the main water supply. Somewhere along this new run a stopcock must be fitted to allow the outside tap to be isolated and drained during cold weather.

Turn off the mains, drain as much as you can from the pipe and cut out a small section so that you can fit a tee joint.

Having fitted the joint turn the water on and check the pipework. If all is well seal the small gap in the wall around the pipe with mastic or filler to prevent insects crawling into your kitchen.

Fitting a washing machine

Plumbing in a washing machine or dishwasher is a straight forward job, ideal for the weekend. It's now easier than ever with time-saving components designed with DIY in mind.

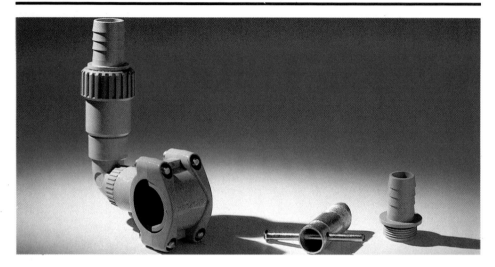

Plumbing in an automatic washing machine or dishwasher is essentially the same job, the only difference being that many washing machines (but not all) require a hot and cold supply whereas most dishwashers need only a single cold feed.

Positioning a washing machine is largely a matter of where it will fit in with your existing arrangements. It need not necessarily go in the kitchen but, for convenience, it seems to be the most popular place. The overriding factor is the drainage. The waste water must be fed into the foul water drains and not into a rainwater gully that soaks away in the garden. A simple check can be made on the suitability of a gully by lifting an adjacent foul drainage manhole cover and pouring water into the gully. If it comes out in the manhole the gully

This washing machine kit contains the parts needed to connect the drainage direct to a plastic sink.

is suitable. Usually it will be the same one that takes the kitchen sink waste.

There are two ways of arranging the waste pipe for a washing machine. One is by using a stand pipe fed into the top of a trap which is then discharged via a 40mm (1½in) pipe out through the wall into the gully. The washing machine manufacturers give the ideal height for the stand pipe in their own installation instructions.

The second way is to use a special washing machine trap that fits on the sink waste in place of the ordinary trap. The new trap has a special spigot to take the flexible washing machine drain hose. After removing the sealed end the machine hose can

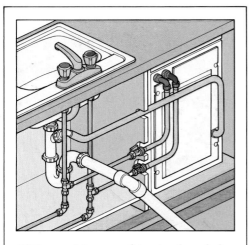

If the washing machine is placed close to the sink, the drainage can be via the sink trap.

be slipped on to the spigot using a drop of washing up liquid as a lubricant. Secure the hose with a jubilee clip to prevent it being pulled off.

The supply pipes to the machine are flexible hoses with ¾in screwed connectors already fitted. The ends with the elbows screw on to the machine fittings – look for colour codes to distinguish the hot from the cold. Each connector must have a rubber washer fitted inside the hose union. Again a drop of washing up liquid on the threads will help overcome resistance when screwing them on – don't be tempted to use grips on these fittings as they are easily damaged. A rag will help you get a good purchase on the knurled nuts.

The hot and cold pipes can be teed off the existing supplies to the sink. Turn off the water and drain the pipes through the kitchen taps. Choose a convenient place to join a compression tee to each pipe and run the 15mm copper supplies to a point near the washing machine. Try to make the connections accessible when the machine is in place.

Fit a pipe clip about 50mm from each pipe end to hold it securely in place – washing machines shut off suddenly and loose pipes may produce a knocking sound. Special lever type washing machine valves should be fitted to the pipe ends to receive the hose connectors. The valves have compression fittings for easy fitting to 15mm copper pipes. With the valves in the closed position – levers at right angles to the valves – test the pipework for leaks.

Before screwing on the connectors flush out any debris from the pipes by opening the valves momentarily with a bucket underneath.

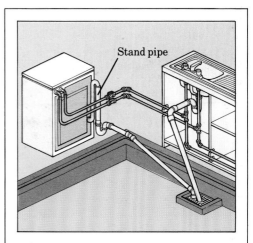

You may need to fit a stand pipe for drainage. Machine manufacturers give instructions as to the ideal height.

Gutters and downpipes

Leaking cast iron gutters and downpipes can cause considerable damage to your home resulting in damp walls and rotting timbers which are expensive to put right. New uPVC rainwater systems are lightweight and easy to fit, requiring no maintenance throughout their long life. So long as you're confident working at heights there is no reason why you shouldn't undertake a complete renewal.

Before you attempt any work from a ladder there are a few common sense safety rules to remember.

Never use a ladder that is broken or not long enough to give you comfortable access. Make sure the ladder is on firm ground and tied to something secure at the top.

● Never overreach from a ladder – move it instead.

Plastic rainwater goods are light, easy to fit and require little maintenance throughout their long life.

● Pitch the ladder at a safe ratio – one foot out from the wall to every four feet of ladder height up.
● Use a ladder stay to keep the head of the ladder away from the eaves.
● Don't struggle moving a ladder on

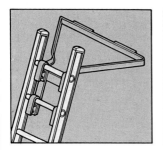 

A ladder stay provides safety and a more comfortable working position.

Always ensure that the ladder foot is securely fixed with ropes or weights.

your own; ask someone to help.
● Remember the rule: one man, one ladder.

Measuring up for new materials can be done entirely from the ground. The height of rainwater pipes can be measured by counting the bricks or measuring the total ceiling heights indoors and adding 300mm (12in) for each floor.

Make a list of all the materials you need before you start removing the old gutters. Stack the new components away from the work area to avoid any damage.

Cast iron gutters can be broken into convenient lengths whilst still fixed to the facia. Use a club hammer to break them into lengths of one metre or so, holding each section to stop it falling. Wear a pair of safety goggles to protect your eyes from flying pieces and check no-one is standing below. The screw fixings can be prised out with a lever bar or claw hammer if they're difficult to undo. Downpipes are usually nailed to the wall through fixing lugs on the sockets. These can be gently levered away

a little at a time. Clear all the old cast iron from the site before commencing the next stage of the job.

Whilst you have unrestricted access to the facias, it's a good idea to give them a couple of coats of paint. Fill all the holes left by the old fixing screws and rub down any loose or flaking areas.

The facia boards will provide a good surface for marking out your falls. The position of the rainwater pipes is determined by the gullies at ground level. Wherever possible, arrange outlets directly above them – use a plumb line to mark the centres of the outlets. Running outlets with two open ends are used for a discharge midway along the run. Stopend outlets are used on the low ends of the runs, ordinary stopends at the high ends.

Work out the falls at an ideal gradient of 1 in 600. Check that facias are level and compensate for those that aren't. The fall of gutters is important; if anything work on too little rather than too much. An excessive fall will allow wind blown rain to

Gutters and downpipes

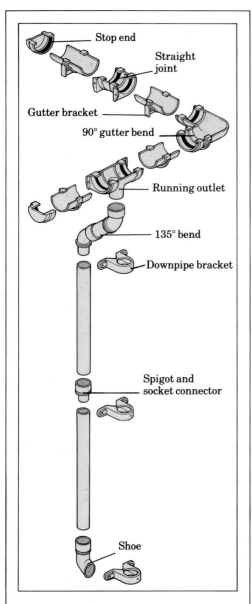

Stop end

Straight joint

Gutter bracket

90° gutter bend

Running outlet

135° bend

Downpipe bracket

Spigot and socket connector

Shoe

If you're replacing a rainwater system, list all the parts you need and measure the length of gutters and downpipes.

run behind the gutter and down the facia as well as leaving an unsightly gap above the gutter. The gutter should be placed just below the tile edge on the high end and fall continuously on the outlet.

Use a string line pulled taut on two brackets at either end of the run and check the fall with a spirit level before fixing the brackets along the string line. Space the brackets not more than one metre apart. A bucket on a hook will hold your screwdriver and bradawl as well as the brackets and fixing screws – much safer than your pockets.

Screw the brackets in position using plated roundhead wood screws providing a 20mm long fixing into the wood. Fix all the angles and outlets to the facia before cutting the gutter; this way you can work out the most economical way of using the odd lengths.

Measure the gutter lengths, taking care to allow for expansion gaps indicated on the fittings. In certain cases you may find it easier to offer the gutter up to the brackets and mark the cutting line rather than struggle with a tape measure.

Use a gutter clip or bracket held over the gutter to give you a square cutting line. A fine tooth wood saw will give the most accurate cut; remove any swarf.

Fix each length of gutter in the brackets and fittings by inserting the rear edge under the lip and snapping the front edge in by pulling down. On cold days the gutters may be made more pliable by storing them indoors for an hour or so. When all the gutters are in place test them with a bucket of

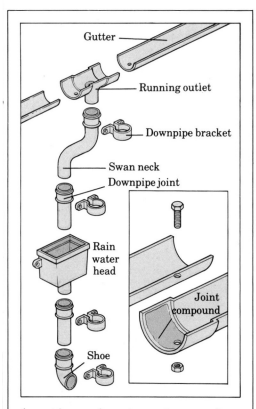

A cast iron rainwater system requires regular painting and maintenance to keep it smart and working efficiently.

at the top of the downpipe and mark out further brackets using a plumb-line. Downpipe brackets should be no further than 2 metres apart; always fit a bracket at the foot of the drop preferably around a rainwater shoe which discharges into the gully. Join lengths of downpipe using spigot/socket connectors.

Bracket fixings are made by drilling and plugging to take roundhead plated screws identical to those used on gutter brackets. If you have to stand the pipe slightly away from the wall you can make small spacers using short lengths of copper tube slipped over the fixing screws. Be sure to add the spacer length to the screw size to maintain a good fixing depth of not less than 20mm.

To prevent fallen leaves blocking the downpipes you can fit small mesh balloons in the top of each gutter outlet.

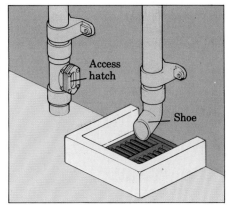

The outlet can be via a shoe into the gully or a direct connection with an access hatch for clearing blockages. Always ensure that the gully grid is kept clear.

water to make certain that the desired fall has been achieved.

The next job is to fit the downpipes. To allow for the difference between the facia and wall you'll need to make up offsets or 'swan necks' using two obtuse (135°) bends and a short off-cut of pipe. This can be done at ground level once you've determined the dimensions. Rainwater pipes simply push fit together, they don't require a sealant. Position a bracket

PICTURE CREDITS

Photography by Malcolm Pendrill:
8, 9, 10, 11, 14, 15, 18, 20, 21,
37TR, 58.
Texas supplied: 30, 31, 35, 36,
37BL, 38, 39, 40, 41, 42, 44, 46, 48,
50.

All illustrated by Ron Haywood
Art Group.

PRINTED IN BELGIUM BY

INTERNATIONAL BOOK PRODUCTION